Rafik Belabbas

Ruminant breeds

Printed by Books on Demand GmbH, Norderstedt / Germany

Rafik Belabbas

Ruminant breeds

Educational book

ScienciaScripts

Imprint
Any brand names and product names mentioned in this book are subject to trademark, brand or patent protection and are trademarks or registered trademarks of their respective holders. The use of brand names, product names, common names, trade names, product descriptions etc. even without a particular marking in this work is in no way to be construed to mean that such names may be regarded as unrestricted in respect of trademark and brand protection legislation and could thus be used by anyone.

Cover image: www.ingimage.com

This book is a translation from the original published under ISBN 978-620-6-70863-6.

Publisher:
Sciencia Scripts
is a trademark of
Dodo Books Indian Ocean Ltd. and OmniScriptum S.R.L publishing group

120 High Road, East Finchley, London, N2 9ED, United Kingdom
Str. Armeneasca 28/1, office 1, Chisinau MD-2012, Republic of Moldova, Europe
Printed at: see last page
ISBN: 978-620-8-06176-0

Foreword

This book describes the origins of ruminant populations and their morphological, phaneroptic and energetic characteristics. It also describes their zootechnical characteristics, enabling us to diagnose breeds and identify animals.

It is divided into 7 chapters: foreign dairy cattle breeds, foreign beef cattle breeds, local cattle breeds, local and foreign sheep breeds and local and foreign goat breeds.

CONTENTS

<u>Introduction</u>

Ethnology derives from Greek words (*ethnos* "people" and *logos* "reason or science"). It is the science that deals with peoples. Its aim is to understand all the characteristics of each ethnic group or race.

Notion of Race...

The race is a subdivision of the species. This word is used to designate a collection of individuals resembling each other, but differing from other individuals of the same species by certain aptitudes or by the harmonious and special development of some of their forms. These individuals have the property of retaining their distinctive characteristics, which they pass on through the generations.

The term "breed" applies mainly to domestic animals, i.e. those bred under the special influence of man; it is not used for animals in the wild (Fauve, 2009).

For individuals to be considered as belonging to the same race, they must (although this is not an absolute condition) resemble each other by the general conformation of the body, or by the special development of certain aptitudes which are also hereditary. These aptitudes are usually combined with identity of form; but it also happens that the general resemblance of conformation, instead of extending to the whole race, is restricted to certain groups within that race (DE weckherlin D'auguste, 1857; Fournier, 2006; Dirand, 2007).

There are different types of breeds:

The breed that has not yet been standardised, or the "traditional population", is present in a given region and is seeking to have its existence officially recognised. This situation is still common in developing countries (e.g. Algeria).

The standard breed (the concept originated in England in the 18th century). This type also includes :
- Large breeds;

- Endangered breeds, or breeds with small numbers.
- International breeds that exist in several regions of the world (*e.g.* the "Holstein" cattle breed).
- Local breeds. A local breed is defined as a breed that is predominantly linked, by its origin, location and breeding method, to a given territory.

The main signs used to distinguish breeds are judged by certain differences in conformation (body, head, bone structure, etc.) and dander (hair, wool, horns, etc.) (Petter, 1987).

I. Dairy cattle breeds

I.1 The Abondance breed : The Abondance breed originated in the Abondance valley in the Chablais massif of Haute-Savoie in France. It is considered to be the 4ème French dairy breed. It is also found in Canada, South America (Chile, Mexico), the Swiss and Italian Alps, Vietnam, and the Near and Middle East (Iraq, Iran and Yemen).

Breed characteristics :

Morphology: convex linear profile. Medium size (eumetric) and proportions are mediolinear. Fine bone structure, fine and dry legs (sign of adaptation to walking).

Table 1: Height and weight of the Abondance breed (www.la-viande.fr).

	Height (cm)	Weight (kg)
Male	150	900-1100
Female	145	600-750

Head: With the exception of the red glasses around the eyes and ears, the bun, belly, leg tips and tail are white.

Horns: the horns are curved forwards, then backwards. The hooves are black.

Coat: Mahogany red piebald (big coat), white head with glasses, mucous membranes and teats vary from pink to brown.

Udder high and well attached; short, fine teats.

Skills :

This is a hardy mountain breed with an exceptionally long lifespan.

An Abondance produces more than 5,800 kg of milk (cheese-making capacity) and a reformed cow provides a carcass of 300 to 350 kg.

Young cattle, and bull calves in particular, grow rapidly and significantly, with the carcass weight of an 18-month-old bull calf being between 320 and 380 kg (Babo, 1998; Baker, 2008; Derville, 2009; Mercier, 2002).

Figure 1: Abondance cow (www.la-viande.fr).

I.2 The Montbéliarde breed: The Montbéliarde is a French breed of cattle born of a cross between indigenous breeds from the Franche-Comté region and a breed from Switzerland. It is considered to be the 2ème French dairy breed in terms of numbers after the Prim'Holstein. It is a renowned milk producer, particularly appreciated for making famous cheeses such as Comté, Morbier, Bleu de Gex and Mont d'Or.

Breed characteristics :

Morphology: convex linear profile. Large format (hypermetric) and mediolinear proportion.

Table 2: Size and weight of the Montbéliarde breed (www.la-viande.fr).

	Height (cm)	Weight (kg)
Male	170	1000-1200
Female	145	650-800

Head: Thin, wide eyes, light mucous membranes and broad muzzle.

Horns: long, thin, white horns set high and curving forward.

Coat: Red piebald with well defined patches, although the head, belly and legs remain white.

Udder: high and wide, set far forward; teats implanted in the centre of the quarters and turned slightly inwards.

Aptitudes : The Montbéliarde is a great dairy cow, but one that retains its beef qualities. The average milk production per cow is over 7000 kg. Its milk is of high cheese-making quality, with a remarkable protein content.

The carcasses of cull cows weigh between 340 and 380 kg, and those of 18-month-old bull calves around 380 kg. This breed is characterised by good longevity (Babo, 1998; Baker, 2008; Derville, 2009; Mercier, 2002).

Figure 2: The Montbéliarde breed (www.la-viande.fr).

I.3. The Brune des Alpes breed : The Brune, previously known as the Brune des Alpes, is a Swiss cattle breed. Bulls of this breed were introduced to Algeria in the 1900s (used for breeding with the local Burne de l'Atlas breed).

Breed characteristics :

Morphology: straight profile. Powerful skeleton.

Table 3: Height and weight of the Brune des Alpes breed.

	Height (cm)	Weight (kg)
Male	155	900-1100
Female	140	650-750

Head: The inside of the ears is hairy and white, reminiscent of plush.

Cornage: lyre- or cup-shaped.

Coat: Mouse-grey or grey-brown with black extremities and mucous membranes, discolouration around the muzzle and inside the ears (white hairs), fading of the colour in the declivities, light stripe, frequent along the topline.

Skills :

The brunette is a good dairy cow that gives nearly 7000 kg of protein-rich milk (fairly good TP/TB ratio: TB = 41.3 ‰ ; TP = 33.6 ‰).

It adapts to all climates, even those in hot regions, thanks to its excellent thermal regulation, which explains its worldwide distribution.

Its beef value is not negligible for a farmer's profitability; a cull cow carcass weighs around 370 kg, while that of an 18-month-old bull calf is around 350 kg (GMQ: 1200 to 1400 g) (Babo, 1998; Baker, 2008; Derville, 2009; Mercier, 2002).

Figure 3: The Alpine Brown breed (www.la-viande.fr).

I.4. The Normande breed: Originating in the Normandy region of France. It is currently found in South America (Mexico, Argentina, Brazil and Ecuador). It is considered to be the 3 French dairy cattle breed in terms of numbers.

Breed characteristics :

Morphology: concave profile. Large format (hypermetric) and mediolinear proportion.

Table 4: Size and weight of the Normande breed (www.la-viande.fr).

	Height (cm)	Weight (kg)
Male	155	900-1200
Female	140	700-800

Head: Rectangular head with convex forehead and wide-set eyes wearing characteristic glasses.

Horns: Crescent-shaped white horns curved forward.

Coat: Tricolored with brindle and white markings, white head with glasses and spotted muzzle.

Udder: well developed and high set.

Skills :

A mixed breed, the Normande is renowned for its dairy and beef qualities.

The cow gives more than 6,500 kg of milk with a high butter content (TB: 44 ‰) and protein (butter-making, cheese-making ability).

Beef conformation: this breed produces heavy, well-conformed, high-quality carcasses; the carcass of a cull cow weighs around 380 kg, that of a 17-month-old steer 370 kg and that of a steer over 400 kg (Babo, 1998; Baker, 2008; Derville, 2009; Mercier, 2002).

Figure 4: The Normande breed (www.la-viande.fr).

I.5. The Prim'Holstein breed: This is considered to be the leading dairy breed in the world, but also in France. A Dutch breed, it was transformed into the FFPN in 1974, after which it was introduced into the USA and Canada, hence its name Frison Holstein, until 1990 when it became Prim'Holstein.

Breed characteristics :

Morphology: Sub-concave profile. Proportion of lanky type with a broad, horizontal pelvis, a very pronounced dairy pattern. A large breed (sub-hypermetric).

Table 5: Height and weight of the Prim'Holstein breed (www.la-viande.fr).

	Height (cm)	Weight (kg)
Male	160	1100
Female	145	700

Head: Long head and broad muzzle.
Horns: fine crescent-shaped horns folded backwards.

Coat: Normally piebald-black with black mucous membrane. The mottling is irregular and highly variable. The scarf and belt pattern is dominant, but any extension of the pattern is possible, with the tips of the legs and tail always remaining white.

Udder: the udder is large and perfectly suited to mechanical milking (attachments, anteroposterior balance).

Skills :

The world's leading dairy breed, with very high quantitative milk performance (record-breaking breed, up to 24,000 kg in one lactation). Extremely high protein content.

While selection for udder shape (udder balance and teat adaptation to mechanical milking) was very effective, it was accompanied by greater susceptibility to intramammary infections (weaker teat sphincter). Mediocre butchery conformation, mediocre yields, precociousness leading to high carcass fatness. Reformed cows producing "bottom of the range" carcasses of 310 - 330 kg.

Selection for high production potential has led to a deterioration in reproductive performance (Babo, 1998; Baker, 2008; Derville, 2009; Mercier, 2002).

Figure 5: The Prim'Holstein breed (www.la-viande.fr).

I.6. The Tarentaise or Tarine breed: The Tarentaise, or Tarine, is a French cattle breed. It is also bred in Italy in the Val d'Aoste under the name Savoiarda. It originated in the Tarentaise Valley in Savoie. It is a great mountain breed, adapted to variations in temperature and rugged terrain (very hardy).

Breed characteristics :

Morphology: straight profile. Size: medium (ellipometric) and Proportion: mediolinear. Thin, solid skeleton.

Table 6: Height and weight of the Tarentaise breed (www.la-viande.fr).

	Height (cm)	Weight (kg)
Male	145	800
Female	130	500

Head: Short head with protruding eye sockets, broad, short muzzle.

Horns: lyre-shaped horns with black tips. Hard, black nails make it an excellent walker.

Coat: uniform reddish fawn, the bull's is darker. Black mucous membranes.

Udder: balanced, firmly attached

Skills :

This hardy breed gives good milk and good meat. A cow provides more than 4500 kg of milk per lactation with a TB of 36 ‰ (a cheese aptitude).

It is easy to fatten, giving it excellent meat potential. The carcass weight of a 17-month-old bullock is around 290 kg.

The Tarentaise lives a long time (good longevity) and calves without complications (Babo, 1998; Baker, 2008; Derville, 2009; Mercier, 2002).

Figure 6: The Tarentaise breed (www.la-viande.fr).

I.7. The Bretonne Pie Noir breed : Originating from the south of Brittany. This is a very old breed, its herd book was created in 1886.

Breed characteristics :

Morphology: small.

Table 7: Height and weight of the Bretonne Pie Noir breed (www.la-viande.fr).

	Height (cm)	Weight (kg)
Male	123	600
Female	117	350-450

Head: fine, expressive head with medium neck.

Horns: Medium-length, lyre-shaped or crescent-shaped horns, white with dark tips.

Coat: Black piebald with clearly defined scarf and belt. White underparts, legs and tuft, black mucous membranes.

Aptitudes: good hardiness, good adaptation to hot climates, good fertility and good longevity. Lactation aptitude: 3590/kg/lactation (Babo, 1998; Baker, 2008; Derville, 2009; Mercier, 2002).

Figure 7: The Bretonne Pie Noir breed (www.la-viande.fr).

I.8. The Salers breed: The breed owes its name to the small town of Salers in the Cantal region of France. A very old breed, used for its cheese and meat, it is considered to be a robust mountain breed (good walker on stony or wet ground).

Breed characteristics :

Morphology: straight profile. Large format (hypermetric). Proportion: elongated (sometimes mediolinear).

Table 8: Height and weight of the Salers breed (www.la-viande.fr).

	Height (cm)	Weight (kg)
Male	153	900-1350
Female	140	800-950

Head: Medium-sized head with light-coloured muzzle.

Horns: long, lyre-shaped horns, old ivory in colour with darker tips. The hooves are black and hard.

Coat: mahogany red without patches, long curly hair, light mucous membranes.

Aptitudes: Suckler breed, quality meat. Hardy, robust, long-lived, easy to calve. The carcass of an 18-month-old bull weighs around 370 kg (Babo, 1998; Baker, 2008; Derville, 2009; Mercier, 2002).

Figure 8: The Salers breed (www.la-viande.fr).

II. Beef cattle breeds

II.1 The Charolais breed: The Charolais is a French breed of cow, originating from the Charolles region in Burgundy, specifically selected for the consumption of their meat, whose individuals are large and of a plain white colour sometimes tending towards cream. It is considered to be the 1ère meat breed in France.

Breed characteristics :

Morphology: Convex linear profile. Large size (hypermetric). Proportion: Sub-revilinear. Skeleton: Strong.

Table 9: Height and weight of the Charolais breed.

	Height (cm)	Weight (kg)
Male	135 - 165	1000 - 1650
Female	135 - 150	700 - 1100

Head: Head small and short, forehead broad. Straight bun, short straight muzzle, broad muzzle, medium sized ears.

Horning: the horns are slightly raised and clear.

Coat: Solid white, sometimes cream, light mucous membranes.

Skills :

This suckler breed has very high growth potential, an excellent feed intake, which means it can put on weight quickly, and excellent beefy conformation.

The carcass weight of an 18-month-old bullock is around 430 kg. The carcass weight of a cull cow is 400 kg. This cow has a high rate of twinning and is hardy and powerful (Babo, 1998; Baker, 2008; Derville, 2009; Mercier, 2002).

Figure 9: Charolais breed (www.la-viande.fr).

II.2 The Limousin breed: The Limousin is a hardy French cattle breed originating from the Limousin region, mainly used for meat production. It is currently found in 80 countries around the world as a purebred or as a crossbreed with local breeds for meat production.

Breed characteristics :

Morphology: Convex profile. Large size (hypermetric) and Proportion: brevilinear. Skeleton: fine. Musculature well developed.

Table 10: Height and weight of the Limousin breed (www.la-viande.fr).

	Height (cm)	Weight (kg)
Male	155	1000 - 1350
Female	135	650 - 850

Head: Short, with broad forehead and muzzle.

Horning: the horns are arched forward.

Coat: Bright to dark wheaten with lighter halos around the eyes and muzzle. The hair is often curly, the mucous membranes are light.

Skills :

The Limousin is a remarkable beef breed, both suckler and hardy. A good growth rate of at least 1000 g/d and good fattening capacity.

The average birth weight is around 42 kg; a 16-month-old bull calf provides a top-of-the-range carcass of around 380 kg, while that of a cull cow reaches the same weight.

This breed has a long lifespan, good fertility and is easy to calve (Babo, 1998; Baker, 2008; Derville, 2009; Mercier, 2002).

Figure 10: The Limousin breed (www.la-viande.fr).

II.3. The Hereford breed: The Hereford breed originated in the region from which it takes its name, the county of Herefordshire (Great Britain). It is a breed that requires little feed and is well adapted to all climates.

Breed characteristics :

Morphology: Convex profile. Medium size (Eumetric) and Proportion: Brevilinear. The trunk is round and massive and the legs are relatively short. Very muscular thighs. Fine bone structure.

Table 11: Size and weight of the Hereford breed (www.la-viande.fr).

	Height (cm)	Weight (kg)
Male	145	900-1100
Female	130	600-800

Horning: in wheeling, some animals have horns, others do not (Polled Hereford).

Coat: Very easy to identify thanks to the particular distribution of colours in the coat: large red mantle; white head, chest, belly and tip of tail. A white line to the withers, white on the legs below the knees and hocks; curly hair.

Abilities: This is the most widespread meat breed in the world. It is hardy, very docile, precocious, fertile, calves easily, has exceptional maternal qualities and lasts a long time.

Calves grow quickly, averaging 40kg at birth, 270kg at weaning at 205 days and 450kg after a year (Babo, 1998; Baker, 2008; Derville, 2009; Mercier, 2002).

Figure 11: The Hereford breed (www.la-viande.fr).

II.4. The Camargue breed: This breed has existed for a very long time in the Camargue region. It is considered to be the only wild European breed (used for fighting).

Breed characteristics :

Morphology: medium format.

Table 12: Size and weight of the Camargue breed (www.la-viande.fr).

	Height (cm)	Weight (kg)
Male	134-140	400-650
Female	115	200-350

Head: broad forehead, depressed.

Cornage: long and powerful, in the shape of a lyre or very raised goblet. The mucous membranes are dark.

Coat: Predominantly glossy black, but can also wear other colours such as brown, grey, white and red.

Aptitudes : good butchery qualities, bright red meat, very expensive and not very fatty.

Figure 12: The Camargue breed (www.la-viande.fr).

III. Local cattle breeds

III.1 BRUNE DE L'ATLAS breed: All types of cattle indigenous to North Africa are known as the Brune de l'Atlas breed. This breed occupies difficult areas, particularly mountainous regions and rangelands. Almost 2/3 of the breed is found in the east of the country.

It is resistant to difficult climatic conditions (heat, cold, drought) and makes good use of poor-quality feed. It is resistant to various parasites and diseases (especially biting insects). The breed does not yet have a Herd Book standard (Amadou, 2019; Yahimi, 2021; Itelv, 2022; Meyer, 2022).

Breed characteristics :

Morphology: Ellipometric format. This breed is brevilinear in all its body parts (to mediolinear). It is a brachycephalic breed. Short neck with little dewlap. Trunk with deep chest. Slightly intricate pelvis, flat thighs and thin buttocks. Thin limbs, black hooves.

Table 13: Size and weight of the Atlas Brown breed (Amadou, 2019).

	Height (cm)	Weight (kg)
Male	115-125	350-400
Female	115-125	250-300

Head: Straight or sub-concave. Head strong, broad and short. Prominent orbital arches.

Cornage: short, crescent-shaped, set slightly on the bun and curving forwards and upwards with black tips. Muzzle surrounded by a border of white hair.

Coat: Dark fawn with black tips, varying from almost blackish brown to reddish brown and dark grey, short hair, brown and slate mucous membranes.

Udder: Although a poor milk producer, the cow has a regular hemispherical udder with small, almost cylindrical teats. Milk production is 4 to 10 litres per day.

Aptitude: The Atlas Brown is a traditional breed (non-specialised, non-improved). It is characterised by poor milk production, with an average birth weight of 20 kg and a GMQ of around 200 g/d. This breed is characterised by its ability to walk on difficult terrain and its good hardiness with regard to climatic conditions and poor feeding (Amadou, 2019; Yahimi, 2021; Itelv, 2022; Meyer, 2022).

The Atlas Brown has undergone changes depending on the environment in which it lives, giving rise to branches that are neither listed nor catalogued. These branches are clearly differentiated from a phenotypic point of view. We can distinguish :

III.1.1 THE EL CHEURFA BREED The El Cheurfa breed: This breed lives on the edge of the forests in the lake and coastal areas of El-Tarf and Annaba where the majority of the population is located. It is present in Jijel and covers the south of Guelma. The coat is light grey, almost whitish, and the muzzle and eyelids are always black.

Figure 13: The El Cheurfa breed (Feliachi, 2003).

III.1.2. The Geulmoise breed : characterised by a dark grey coat, living in forested areas in the Guelma and Jijel regions, this population makes up the majority of the stock. The horns are short, thick at the front and then narrow, lyre-shaped in females and crescent-shaped towards the front in males. Small size: height at withers: 80 to 110 cm. Weight: 175 - 200 kg for females and 220 - 250 kg for males (Amadou, 2019; Yahimi, 2021; Itelv, 2022; Meyer, 2022).

Figure 14: The Geulmoise breed (Feliachi, 2003).

III.1.3. The Sétifienne breed: With a uniform blackish coat, it has a good conformation. Its size and weight vary according to the region where it lives. The tail is black, long and sometimes drags along the ground. The brown line down the back is characteristic of this population. The weight of females in the cereal-growing highlands is close to that of imported females. Milk production can reach 1500 kg/year. It is located in the Babors mountains (Amadou, 2019; Yahimi, 2021; Itelv, 2022; Meyer, 2022).

Figure 15: The Setifienne breed (Feliachi, 2003).

III.1.4. The Chélifienne breed: originating from the Chélif region. It is very small in size. It is characterised by a fawn coat, a short head, hooked horns, protruding eye sockets surrounded by dark brown glasses and a long black tail that touches the ground. It is found in the Dahra mountains (Amadou, 2019; Yahimi, 2021; Itelv, 2022; Meyer, 2022).

Figure 16: The Chélifienne breed (Feliachi, 2003).

III.1.5. The Djerba breed: This breed comes from the Biskra region and is characterised by a dark brown coat, a narrow head, a rounded croup and a long tail. The very small size, adapted to the very difficult environment of the South.

III.1.6. The Kabyle and Chaouia breeds: derived respectively from the Guelmoise and Cheurfa breeds following successive changes in cattle breeding. It is found in Kabylia and the Aurès (Amadou, 2019; Yahimi, 2021; Itelv, 2022; Meyer, 2022).

IV. Foreign sheep breeds

IV.1. The Lacaune breed :

Lacaune is a French sheep breed. It is the leading French breed in terms of numbers. Originally a mixed breed, it now comprises two varieties: Lacaune milk and Lacaune meat. The milk Lacaune is mainly farmed in the Roquefort area for milk production, mainly for cheese processing under this designation of origin.

The Lacaune viande, a meat animal, is farmed over a wider geographical area, under a strict suckler system for the main production of fattened lambs for slaughter.

It is exported to many countries including Portugal, Spain, Greece, Tunisia, Slovakia, Switzerland, Germany, Austria, Hungary and Brazil.

Breed characteristics :

Origin: France.

Morphology: Relatively hypermetric (large). Broad, long body, deep chest, long, docked tail. Legs of medium length, proportionate and well poised.

Table 14: Height and weight of the Lacaune breed (www.la-viande.fr).

	Height (cm)	Weight (kg)
Aries	70 à 80	100
Ewes	70	70

Head : Slightly arched profile. The head is triangular, fine, covered with fine white hair with a silvery tinge. Forehead slightly domed, long horizontal ears.

Horns: hornless in both sexes.

Fleece: white; head, neck and belly bare (non-invasive); fleece weighs less than 2.5 kg.

Aptitude: mixed breed (milk and meat). Milk production between 100 and 120 L and up to 350 L/180 days (TP: 52‰, TB: 72 ‰). Prolificity is 175% (Babo, 2000; Gillespie, 2009; Jussiau, 2013; Vaissaire, 2014; Wikipedia, 2022).

Figure 17: The Lacaune breed (www.la-viande.fr).

IV.2 The Ile de France race :

This is a breed of sheep selected from 1840, near Paris, from crosses between Rambouillet Merino ewes and Dishley rams imported from England. It has a good balance of beefy and maternal qualities.

Breed characteristics :

Origin: France.

Morphology: heavy breed (hypermetric). Long and ample trunk, short neck without dewlap, fold or tie. Short limbs; regular legs; well developed, deep legs.

Table 15: Size and weight of the Ile de France breed (www.la-viande.fr).

	Height (cm)	Weight (kg)
Aries	78	110 à 150
Ewes	70	70 à 90

Head : profile almost straight, slightly arched in the male. The head is strong, wider at the skull, ears large, standing horizontally; wrinkles on the nose.

Horns: hornless.

Fleece: extensive white covering the top of the head, the ganaches, the rear edge of the cheeks, the front legs up to the knees (invasive). It is closed and compacted, with square locks of good length, weighing around 5 to 6 kg for a ram and 4 kg for a ewe; fine wool.

Aptitude: the breed is bred for its wool, meat production and also for cross-breeding. The breed is precocious, prolific (prolificacy > 170%), a good milk producer and very well conformed; the lambs have a high growth rate and their meat is of good quality; average GMQ (10- 70 D) is 350-400g in males (Babo, 2000; Gillespie, 2009; Jussiau, 2013; Vaissaire, 2014; Wikipedia, 2022).

Figure 18: The Ile de France breed (www.la-viande.fr).

IV.3. The Charmoise breed :

It is mainly found in the Centre-West (Vienne) and in the South-West of France (Ardennes). It makes good use of forage subtypes and adapts particularly well to difficult environments.

This breed takes its name from the Charmoise farm near Pontlevoy in the Loir-et-Cher region, where it was bred between 1838 and 1852 by an agronomist and breeder, Édouard Malingié.

Breed characteristics :

Origin: France.

Morphology: Medium-sized breed. The body has a broad chest, level back, wide croup and a clearly rounded, well let down leg. The legs are short, slender and wide apart.

Table 16: Height and weight of the Charmoise breed (www.la-viande.fr).

	Height (cm)	Weight (kg)
Aries	65-70	80 à 90
Ewes	65	55 à 70

Head: Straight profile. Mucous membranes light, pinkish, almost brown. Eyes exorbites. Ears erect.

Horns: hornless.

Fleece: fine, white wool with bare head and limbs (semi-invasive fleece).

Aptitude: hardy pasture breed. One of the breed's great assets is its natural deseasoning and excellent butchering ability. The prolificacy rate is 120% (Babo, 2000; Gillespie, 2009; Jussiau, 2013; Vaissaire, 2014; Wikipedia, 2022).

Figure 19: The Charmoise breed (www.la-viande.fr).

IV.4. The Texel breed :

The Texel is a breed of sheep originating on the island of the same name in the Netherlands. It is a very old breed that has been improved over time. The Texel is a sheep with dense, fairly long white wool, from which emerges a naked head with a black nose.

Breed characteristics :

Origin: Netherlands (Texel Island).

Morphology: relatively heavy breed. Conical trunk, thick withers, horizontal, square croup, broad pelvis. The legs are strong and end in black hooves.

Table 17: Height and weight of the Texel breed (www.la-viande.fr).

	Height (cm)	Weight (kg)
Aries	70-80	110 à 140
Ewes	70	70 à 90

Head: Straight profile. The head is characteristic with the top of the skull quite flat, the muzzle broad and short and the ears thick and slightly erect. The nose is dark (often black).

Horns: hornless.

Fleece: white, abundant and thick to the point of blurring the shape of the body, not covering the head and limbs (semi-invasive); weight of the fleece in males: 6-8 kg and in females: 4.5 kg.

Aptitude: meat and wool; high prolificacy (175-200%); good milk production; rapid growth rate (a single male can gain 400 to 500 g per day between 10 and 30 days; at 70 days, a single male weighs 30 kg). The breed's beef qualities are conformation and not too much fat (Babo, 2000; Gillespie, 2009; Jussiau, 2013; Vaissaire, 2014; Wikipedia, 2022).

Figure 20: The Texel breed (www.la-viande.fr).

IV.5. The Suffolk breed :

Originally from the United Kingdom. It is the result of crosses between Norfolk ewes and Southdown.

Breed characteristics :

Origin: England.

Morphology: Large breed. Body long, powerful and cylindrical. Medium neck, broad, compact shoulders, flat, long back, rounded ribs, deep chest and fairly high set tail.

Correct legs, wide apart with black hair at the ends.

Table 18: Height and weight of the Suffolk breed (www.la-viande.fr).

	Height (cm)	Weight (kg)
Aries	/	90 à 150
Ewes	/	70 à 90

Head: Straight, slightly convex. Black hair; ears long and horizontal in the female and drooping in the male. Head devoid of wool and covered with fine, shiny hair .

Horns: hornless.

Coat: the skin is black and the wool is white, covering neither the head, nor the ears, nor the lower half of the limbs (semi-invasive); the weight of a shearing remains low.

Aptitude: meat breed. The breed is hardy, precocious and prolific (prolificacy: 130 - 180%), with good milk production (Babo, 2000; Gillespie, 2009; Jussiau, 2013; Vaissaire, 2014; Wikipedia, 2022).

Figure 21: The Suffolk breed (www.la-viande.fr).

IV.6. The South Down breed :

The Southdown is a breed of sheep originating in the south of England. It is used as a purebred or to improve French breeds through cross-breeding.

Breed characteristics :

Origin: England.

Morphology: Relatively large (hypermetric). Short neck, broad back, remarkable butchery conformation. Legs well turned.

Table 19: Height and weight of the South Down breed (www.la-viande.fr).

	Height (cm)	Weight (kg)
Aries	/	90 - 120
Ewes	/	60 - 80

Head: Concavilinear. Broad head, short muzzle, grey-brown face, sometimes blackish nose and black nasal mucous membrane; small, very fine, erect ears.

Horns: hornless.

Fleece: white, very thick (forehead, cheeks full); a ram's fleece weighs between 5 and 4.5 kg, a ewe's between 2 and 2.5 kg.

Aptitude: this breed is renowned for the very good conformation of its lambs (very large hindquarters, well rounded legs; prolificacy rate of 170%) (Babo, 2000; Gillespie, 2009; Jussiau, 2013; Vaissaire, 2014; Wikipedia, 2022).

Figure 22: The South Down breed (www.la-viande.fr).

IV.7. The Bukhara Romanov breed :

A Russian breed of sheep. It is a black sheep with more or less grey wool. It is one of the short-tailed sheep breeds of Northern Europe.

Breed characteristics :

Origin: Russia.

Morphology: Medium-sized breed. Long body and legs. The ribs are well rounded and the animal appears a little high-legged.

Table 20: Height and weight of the South Down breed (www.la-viande.fr).

	Height (cm)	Weight (kg)
Aries	/	70 - 90
Ewes	/	60 - 70

Head: Bracing profile. Head small, angular, black skinned, with white patches on forehead and muzzle, erect ears. The eyes are large.

Horns: No horns.

Fleece: black at birth then becomes grey-blue (semi- to non-invasive).

Aptitude: Sexual precocity. Prolificity: 260-320%. Sexual activity all year round. Very good dairy (Babo, 2000; Gillespie, 2009; Jussiau, 2013; Vaissaire, 2014; Wikipedia, 2022).

Figure 23: The South Down breed (www.la-viande.fr).

IV.8. The Merino breed (Mérinos de Rambouillet):

Merinos are a breed of sheep native to Spain bred primarily for their wool. The term merino is also used to designate the very fine, comfortable wool produced from the dense, curly fleece of these sheep. This wool, produced mainly in Australia, is used in the production of high quality woollens.

Breed characteristics :

Origin: Spain.

Morphology: medium (eumetric). Cylindrical body with broad chest and horizontal back. The neck is short with a slightly developed dewlap. Broad withers. Tail long and fine. The legs are sturdy and the legs are fairly thick, set wide apart and plumb. The leg is rather thick; strong skeleton.

Table 21: Size and weight of the Merino breed (Mérinos de Rambouillet) (www.la-viande.fr).

	Height (cm)	Weight (kg)
Aries	/	70 - 90
Ewes	/	60 - 70

Head : Straight profile, slightly arched. The head is fine and short; the ears are short and horizontal. The nose is topped by several folds.

Horns: only the male has regularly spiralling horns.

Fleece: The fleece is very invasive and very tight with square locks. The wool is white, extra fine (15-22 microns) and very wavy (length: 6-7 cm). Fleece weight: 10% of live weight (Belier: 5-8 kg).

Aptitude: great wool qualities. Its white wool is abundant, fine, elastic and resistant. It is also a hardy breed that adapts easily to all climates and accepts all diets; weak growth, mediocre butchery conformation; low prolificacy (130%) (Babo, 2000; Gillespie, 2009; Jussiau, 2013; Vaissaire, 2014; Wikipedia, 2022).

Figure 24: The Merino breed (Mérinos de Rambouillet) (www.la-viande.fr).

V. Algerian sheep breeds

There are no fewer than 900 breeds of sheep in the world. Each breed has its own external characteristics and character traits. Algerian sheep are extremely diverse. Classification is based on the existence of three main breeds (Ouled Djellal, Hamra and the Rumbi breed) and secondary breeds.

V.1 The main breeds of sheep :

V.1.1. The Ouled Djellal race (the white Arab race) :

This is the true steppe sheep, best suited to nomadic life. The Ouled Djellal breed is found in central and eastern Algeria, a vast area stretching from Oued Touil (Laghouat/Ksar Chellala) to the Tunisian border. It is the best meat breed in Algeria and is now tending to replace certain breeds in their own cradle, such as the El Hamra breed.

Breed characteristics :

Cradle of the breed: central and eastern Algeria, a vast area stretching from Oued Touil (Laghouat/Ksar Chellala) to the Tunisian border.

Morphology : Well proportioned, high waist, slightly narrow chest, flat ribs and legs and long, strong, walking legs. The tail is fine and of medium length.

Head: White with medium sized hanging ears set high on the head. There is a slight depression at the base of the nose.

Horns: spiral-shaped, medium length in rams and absent in ewes.

Colour: White throughout. However, some "safra ewes" have a light straw colour.

Wool: The wool is fine and covers the whole body up to the knees and hocks for the Hodna and Chellala varieties, while the belly and underside of the neck are bare for most Ouled-Djellal sheep. This breed fears extreme cold (Boushaba, 2007; Lakhdari, 2015; Kebbab, 2018).

Figure 25: The Ouled - Djellal breed (www.algerlablanche.com).

The Ouled Djellal breed comprises three varieties:

The Ouled Djellal or Djellalia variety: This variety represents 16% of the Ouled Djellal breed. It is found in the Zibans region (Biskra and Touggourt).

It is a long, leggy sheep, adapted to nomadic life. The skeleton is very fine, the leg long and flat, and the meat has a slight oozing flavour.

The wool is white, fine and garlicky, the belly and underside of the neck are bare, the horns are medium-sized, spiral-shaped and may be present in ewes.

This variety makes excellent use of rangelands. It is the sheep of the nomadic tribes of the southern foothills of the Saharan Atlas (Boushaba, 2007; Lakhdari, 2015).

Table 22: Size and weight of the Djellalia variety (Lakhdari, 2015; Kebbab, 2018).

	Height (cm)	Weight (kg)
Aries	80	68
Ewes	70	48

Figure 26: The Djellalia variety (www.algerlablanche.com).

The Ouled Naïl variety:

Commonly known as Hodnia, it accounts for 70% of the Ouled Djellal population and is the heaviest type. It is found in the Hodna, Sidi Aissa, M'sila, Barika and Sétif regions. The sheep is well proportioned and tall. They are light straw or white in colour. The wool covers the entire body right down to the hock (Boushaba, 2007; Lakhdari, 2015; Kebbab, 2018).

Table 23: Size and weight of the Hodnia or Ouled Naïl variety (Lakhdari, 2015).

	Height (cm)	Weight (kg)
Aries	82	82
Ewes	74	57

Figure 27: The Hodnia or Ouled Naïl variety (www.algerlablanche.com).

The Chellala or Chellalia variety: also known as the Taadmit breed. It is found in the region of Laghouat, Chellala, Taguine (Oued Touil) and Bokhari. This variety makes up 5 to 10% of the Ouled Djellal population.

This variety is the smallest in size and has very fine wool. Rams of this type are considered less combative than those of the Ouled Djellal variety and are often cloddy (hornless) (Boushaba, 2007; Lakhdari, 2015; Kebbab, 2018).

Table 24: Size and weight of the Chellala or Chellalia variety (Lakhdari, 2015).

	Height (cm)	Weight (kg)
Aries	75	73
Ewes	70	47

Figure 28: The Chellala or Chellalia variety (www.algerlablanche.com).

V.1.2. The Hamra or Beni Ighil breed :

It is the second most important breed in Algeria (representing 22% of the Algerian sheep population). It is a Berber breed originating from the high plains of the west (Saïda, Mécheria, Ain-Safra and El Aricha in the Wilaya of Tlemecen). It is the best meat breed (threatened with extinction).

Breed characteristics :

Cradle of the breed: In Algeria: from the Chott Chergui to the Moroccan border. In Morocco: the Moroccan Eastern High Atlas.

Morphology: reduced (ellipometric). Body small but short, stocky and broad, leg short and round, skeleton fine. The ears are medium sized and drooping and the tail is thin and of medium length.

Head: convex profile, with a convex chamfer.

Horns: Medium-sized and spiral-shaped.

Colour: brown skin, black mucous membranes, brown head and legs, dark mahogany red, almost black. The fleece is white and mottled reddish-brown.

Quality: The quality of its meat is excellent and it is considered to be one of the best meat breeds in Algeria because of its fine bones and round lines (Boushaba, 2007; Lakhdari, 2015; Kebbab, 2018).

Table 25: Size and weight of the Hamra sheep breed (Lakhdari, 2015).

	Height (cm)	Weight (kg)
Aries	76	71
Ewes	67	40

Figure 29: The Hamra breed of sheep (www.algerlablanche.com).

V.1.3. The Rumbi breed :

The Rembi breed has the same characteristics as the Ouled-Djellal breed, except for the colour of its limbs and head, which is fawn. It accounts for 11% of the national herd.

Breed characteristics :

Cradle of the breed: The cradle of the breed stretches from Oued Touil in the east to Chott Chergui in the west.

Morphology: Good conformation. It is a particularly hardy and productive breed with a massive skeleton, very sturdy legs resembling mouflon and very hard hooves. The tail is thin and of medium length.

Head: Bracing profile. Reddish-brown head. Medium length floppy ears.

Horns: spiral and massive.

Colour: chamois. The wool covers the whole body down to the knees and hocks.

Quality: hardy and highly productive breed. It is used for meat production (Boushaba, 2007; Lakhdari, 2015; Kebbab, 2018).

Table 26: Size and weight of the Rumbi breed (Lakhdari, 2015).

	Height (cm)	Weight (kg)
Aries	77	80
Ewes	71	62

Figure 30: The Rumbi breed (www.algerlablanche.com).

V.2 Secondary breeds of sheep :

V.2.1. Man's or Men's breed :

It represents 0.19% of the national sheep population. It is a very hardy palm tree animal that tolerates Saharan conditions very well, and is often referred to as the Tafilalet breed.

Breed characteristics :

Cradle of the breed: Oases in south-western Algeria (Gourara, Touat, Tidikelt).

Morphology: small size. Very fine skeleton, high on legs, well developed belly and high prolificacy. The neck is long, thin and pendulous in ewes, rarely in rams. The tail is thin and long with white tips.

Table 27: Man or Men breed size and weight (Lakhdari, 2015).

	Height (cm)	Weight (kg)
Aries	75	46
Ewes	60	37

Head: profile is rounded. This breed is also characterised by a fine head and large, pendulous ears.

Horns: Neither ewes nor rams have horns. The absence of horns in males distinguishes the D'man breed from other North African breeds.

Colour: Variable pigmentation, the head and fleece may be entirely black, brown or white, or a juxtaposition of 2 or 3 colours.

Fleece: small and black or dark brown. The belly, chest and legs are without wool, sometimes the fleece covers only the back.

Quality: hardy, adapted to Saharan conditions, very prolific. The meat is mediocre (tough and difficult to chew) (Boushaba, 2007; Lakhdari, 2015; Kebbab, 2018).

Figure 31: The Man or Men breed (www.algerlablanche.com).

V.2.2. The Barbarine breed :

It represents 0.27% of the national herd. The breed is related to the Middle Eastern barbary and the Asian barbary. The fat reserve in the tail and its large hooves make it a breed adapted to the conditions of the Eastern Erg, its main habitat.

Breed characteristics :

Cradle of the breed: the Oued Souf fat-tailed sheep is found in the Erg Oriental (Oued Souf) as far as the Tunisian border.

Morphology: Good conformation, compact body, short neck, short legs, broad, deep chest. Big tail 1 to 2 kg, after fattening 3 to 4 kg. Its large hooves make it an excellent walker in the dunes of the Sahara (Souf).

Table 28: Size and weight of the Barbary breed (Lakhdari, 2015).

	Height (cm)	Weight (kg)
Aries	70	45
Ewes	64	37

Head: profile rounded. Ears medium, hanging.

Horns: Developed in males, absent in females.

Colour: the body is white except for the head and legs, which may be brown or black.

Quality: The quality of the meat is good, but not appreciated in Algeria because of its large tail and smell (Boushaba, 2007; Lakhdari, 2015; Kebbab, 2018).

Figure 32: The Barbarian breed (www.algerlablanche.com).

V.2.3. The Berber race :

It is the $2^{\text{ème}}$ most important breed, accounting for 25% of the national sheep population. The Berber, considered to be the oldest Algerian breed, is traditionally reared in the mountainous massifs of northern Algeria.

Breed characteristics :

Cradle of the breed: This is a breed from the Tell Mountains (Tellian Atlas in North Africa).

Morphology: Small in size. Body small but short, stocky and broad, leg short and round, skeleton fine. Tail fine, medium length.

Table 29: Size and weight of the Berber breed (Lakhdari, 2015).

	Height (cm)	Weight (kg)
Aries	65	45
Ewes	60	35

Head : Convex, arched. Ears medium sized and hanging.

Horns: spiral, medium.

Colour: the skin is brown, the mucous membranes black, the head and legs are brown, dark red, almost black. The wool is white with a reddish-brown ruff.

Fleece: White, wiry, shiny wool known as Azoulai, with some specimens spotted black.

Quality: The meat is of average quality (a little tough). The legs are long and flat, with little development. It is a very hardy animal, tolerating the extreme cold of the mountains and making good use of scrubby mountain pastures (Boushaba, 2007; Lakhdari, 2015; Kebbab, 2018).

Figure 33: The Berber race (www.algerlablanche.com).

V.2.4. The Sidahou or Targuia breed :

The Sidahou, also known as the Targui, is a breed that originated in Mali and is mainly farmed by the Tuareg, but is also found in the Sahara. It accounts for around 0.13% of Algerian sheep. It is the only Algerian breed without wool, but with a body covered in hair resembling a goat. This breed is resistant to the Saharan climate and long marches, and is the only breed that can live on the extensive pastures of the Sahara.

Breed characteristics :

Cradle of the breed: This breed is found in the great Algerian Sahara from Bechar via Adrar to Janet.

Morphology: The chest is narrow and the legs are long and high, making it suitable for walking long distances (up to 1000 kilometres). The tail is slender, very long, almost at ground level, with a white tip.

Table 30: Size and weight of the Sidahou or Targuia breed (Lakhdari, 2015).

	Height (cm)	Weight (kg)
Aries	77	41
Ewes	76	37

Head : The muzzle is very curved, the ears are large and hanging.
Horns: absent or small and curved in males.

Colour: The colour of the hair is black or straw or mixed.

Fleece: This is the only Algerian breed without wool but with a body covered in hair. The Targuia resembles a goat except that it has a long tail and a sheep-like bleat.

Quality: Targuia meat is below average and hard to chew. The short, flat leg and shoulder are not supplied with meat (Boushaba, 2007; Lakhdari, 2015; Kebbab, 2018).

Figure 34: The Sidahou or Targuia breed (www.algerlablanche.com).

VI. Foreign breeds of goats

VI.1 The Alpine breed :

The Alpine goat is native to the Alps of France and Switzerland. It is a medium-sized goat with a buff-coloured coat, which is the most widespread. This breed is an excellent milk producer.

Breed characteristics :

Head : Concave profile. The head is triangular. Broad forehead and muzzle, with or without tassels (outgrowths of skin behind the throat) and with a goatee. The eyes are prominent and the ears are carried erect forward, in a rather firm cone.

Morphology: Medium-sized breed (eumetrical), free neck, straight topline, broad, slightly sloping croup, deep chest and broad pelvis. The legs are solid.

Table 31: Height and weight of the Alpine breed (www.wikipedia.org).

	Height (cm)	Weight (kg)
Male	90 - 100	80 - 100
Female	50 - 80	50 - 70

Horned: with or without horns.

Coat: The coat comes in various colour patterns: fawn (buff) or brown (pain brulé) are the most common, with black legs and dorsal stripe.

Udder: voluminous, with fine, supple skin, well attached at the front and back. Retracts well after milking. The teats are well detached from the udder and point forward, making the Alpine perfectly suited to machine milking.

Aptitudes : The Alpine goat is an excellent milk producer. Average production is 790 kg per 272-day lactation with an average TB of 37‰ and TP of 32.5‰. Females are very precocious and fertile (Fournier, 2006; Vaissaire, 2014; Wikipedia, 2022).

Figure 35: The Alpine goat (www.wikipedia.org).

VI.2 The Saanen breed :

The Saanen or Blanche de Gessenay is a goat breed originating in the Sannenland and Obersimmental regions of Switzerland. White and cream are the only accepted coat colours. It is a strong milk producer.

Breed characteristics :

Head: profile almost straight to sub-concave. Forehead and muzzle broad, with or without tassels, sometimes with a goatee. Ears carried horizontally.

Morphology: A large breed. The body is deep, thick and well boned, the chest deep, broad and long (large thoracic capacity). Legs strong and straight.

Table 32: Height and weight of the Saanen breed (www.wikipedia.org).

	Height (cm)	Weight (kg)
Male	90 - 100	90 - 120
Female	70 - 80	70 - 80

Horned: with or without horns.

Coat: Short, dense and silky. Uniformly white.

Udder: globular, very wide at the top, which gives it greater width than depth.

Aptitudes: an excellent milk producer in the world, 800 to 1000 kg in lactation from 270 to 300 days whose average TB is 35‰ and TP is 31‰. Females give excellent kids whose meat is very appreciable (Fournier, 2006; Vaissaire, 2014; Wikipedia, 2022).

Figure 36: Saanen goat (www.wikipedia.org).

VI.3. The Poitevin breed :

A medium-sized goat breed originating in central-western France (Poitou).

Breed characteristics :

Head: Straight profile. The head is black and triangular with two small white patches which sometimes extend to very pronounced white stripes on either side of the bridge of nose, the forehead and the bun are fairly straight.

Morphology : The body is voluminous, with a long, straight back and deep chest. Long, supple neck, proud head carriage. Strong bones.

Table 33: Size and weight of the Poitevine breed.

	Height (cm)	Weight (kg)
Male	60 - 80	55 - 75
Female	60 - 70	40 - 65

Horned: with or without horns.

Coat: Generally brown, varying shades of dark to black. White is found on the belly, the inside of the legs and the underside of the tail.

Udder: The udder is elongated and regular, with supple skin. The Poitevin goat may have a goatee and tassels, but this is not the case for all animals.

Aptitudes: hardy, good dairy: lactation: 500 - 1000 kg (Fournier, 2006; Vaissaire, 2014; Wikipedia, 2022).

Figure 37: The Poitevin goat (www.wikipedia.org).

VI.4. The Murciano-Granadina breed :

The Murciano-Granadina is a Spanish dairy goat breed. It was created in 1975 by crossing two breeds, the mahogany-coloured Murciana de Murcia and the black Granadina de Granada. The Murciana de Murcia breed has been imported into Algeria since colonial times.

Breed characteristics :

Head : Concave profile. The head is fine and the ears carried horizontally.

Morphology: Medium-sized breed. The neck is long, the body is long and rounded and the tail is erect. The neck is slender, the body elongated, the chest broad and the abdomen voluminous, the sacrum convex. The tail is short and straight. The legs are of medium height, strong and slightly twisted, as they cover the voluminous udder.

Table 34: Size and weight of the Murciano-Granadina breed (www.wikipedia.org).

	Height (cm)	Weight (kg)
Male	70	50 - 60
Female	77	30 - 50

Horned: hornless (rare).

Coat: Solid colour, generally mahogany, sometimes black. The hair is short and coarse, although it is longer in males than in females.

Udder: The teats of the udder face forwards and outwards, the skin is thin and hairless.

Aptitudes: a hardy goat with good dairy qualities. Milk production of around 500 litres in a lactation of 210 days on average (average TB is 56‰ and TP is 36 ‰). In Spain, its milk is mainly used for cheese production. They are also used for meat, not least because of their rapid growth. These goats are deseasoned with a prolificacy of around 200% (Fournier, 2006; Vaissaire, 2014; Wikipedia, 2022).

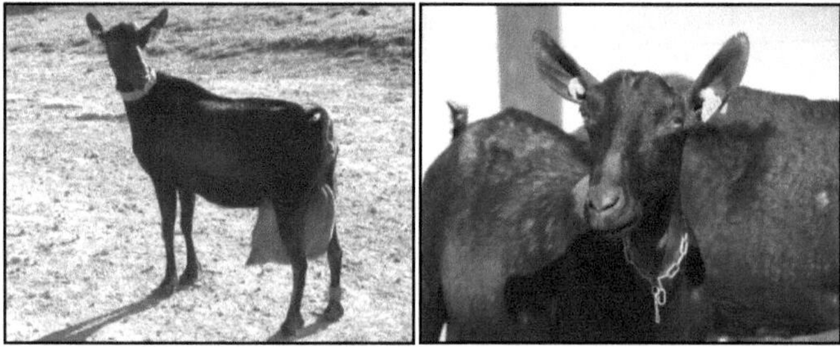

Figure 38: The Murciano-Granadina goat (www.wikipedia.org).

VI.5. The Angora breed :

The Angora goat, also known as the Tibetan goat, is a breed of goat that originated in the Himalayas (Kashmir and Tibet) and was later introduced to Turkey and Asia Minor.

Breed characteristics :

Head: Straight profile. The head is fine. Hanging ears carried forward.
Morphology: reduced format.

Table 35: Height and weight of the Angora breed (www.wikipedia.org).

	Height (cm)	Weight (kg)
Male	60 - 80	40 - 70
Female	60	30 - 50

Horning: presence of horns in males

Coat: Long, fine, silky white hair used to make a knitting wool called "mahair". Length: 8-12cm, fineness: 26-30 microns. No jars. The wool is white, the fleece curly or frizzy.

Aptitudes: It is hardy and has a good wool yield (5kg of mohair/year/sire). It produces little meat and especially little milk (Fournier, 2006; Vaissaire, 2014; Wikipedia, 2022).

Figure 39: Angora goats (www.wikipedia.org).

VII. Algerian goat breeds

The *Capra hircus* species is found in Algeria in the form of a mosaic of highly varied populations, all belonging to traditional populations. In addition to these local populations, which generally have Nubian blood, it includes animals with mixed blood from standardised breeds.

Algeria's goat population is made up of four major types: **the Arbia breed**, mainly found in the **Laghouat** region; the **Kabyle** breed, found in the **Kabylie and Aurès mountains**; the **Makatia** breed, found on the **high plateaux** and in certain **areas of the north**; and finally the **M'Zabia** breed, found in the **northern** part **of the Sahara**.

VII.1 The local population :

VII.1.1. The Arabian goat (ARABIA or ARBIA) :

This is the most dominant population, related to the Nubian race, and is found mainly in the high plateaux, steppe and semi-steppe areas.

Breed characteristics :

Morphology: reduced format (ellipometric).

Table 36: Size and weight of the Arbia breed (Laouadi, 2020).

	Height (cm)	Weight (kg)
Male	70 - 74	50 - 60
Female	50 - 64	32 - 35

Head : Straight profile. Bridge of nose usually straight (slightly convex in some individuals). Ears long, broad and hanging.

Horns: Generally hornless, but when they do exist (especially in males), they are moderately long and point backwards.

Coat: polychrome (multicoloured) coat: black, grey and brown with long hair (10-17 cm). Often black with white legs. The head is a solid colour or has two white markings on either side.

Ability: The Arabian goat has a mediocre milk production (average of between 0.25 and 0.75 litres per day). It is bred for the meat of kids but also for milk (Boubekeur, 2010; Laouadi, 2020, Nessah, 2020, Wikipedia, 2022).

Figure 40: The Arbia goat (Laouadi, 2020).

We have two types: sedentary and transhumant.

a. Sedentary type :

Morphology: The average height is 70cm for the male and 63cm for the female, while their respective weights are 50kg and 35kg. The body is elongated with a straight topline and straight muzzle. The ears are fairly long (17 cm).

Horned: Medium-length horns pointing backwards.

Coat: The coat is long, from 10 to 17 cm, and polychrome white, black piebald, and brown. The head is either a solid colour or with markings.

Aptitude: milk production is 0.5 litres per day (Boubekeur, 2010; Laouadi, 2020, Nessah, 2020, Wikipedia, 2022).

b.Transhumant type :

Morphology: the average height is 74 cm for the male and 64 cm for the female, with respective weights of 60 kg and 32 kg. The body is elongated, straight on top, but convex in some individuals. The ears are very large.

Horns: The head has fairly long horns pointing backwards (especially in males).
Coat: Long hair 14 to 21cm long, predominantly piebald black.

Aptitude: milk production is 0.25-0.75 litres per day (Boubekeur, 2010; Laouadi, 2020, Nessah, 2020, Wikipedia, 2022).

VII.1.2. The MAKATIA goat :

It originates from Ouled Nail. It is found in the Laghouat region. It is undoubtedly the result of a cross between the ARABIA and the CHERKIA, and is generally kept in association with the sedentary ARABIA goat.

Breed characteristics :

Morphology: Medium size. The body is elongated with a straight topline.

Head: Convex profile. Strong head in males and females. Goatee and two pendants (less frequent). Ears long (16cm) and drooping.

Table 37: Size and weight of the Makatia breed (Laouadi, 2020).

	Height (cm)	Weight (kg)
Male	72	60
Female	63	40

Horning: horns pointing backwards.

Coat: varied grey, beige, white and brown with short, fine hair, 3-5 cm long.

Udder: The udder is well balanced, square, high and well attached, and 60% of females have large teats.

Ability: milk production is 1 to 2 litres per day (Boubekeur, 2010; Laouadi, 2020, Nessah, 2020, Wikipedia, 2022).

Figure 41: The Makatia goat (Itelv, 2015).

VII.1.3. The Kabyle goat (Naine de Kabylie) :

This is a native goat that inhabits the mountain ranges of Kabylia and the Aurès.

Breed characteristics :

Morphology: reduced size (ellipometric). Robust, massive, elongated body with straight topline.

Head : The head is fine. The ears are small and pointed for white coated dogs and medium long for beige coated dogs.

Table 38: Size and weight of the Naine de Kabylie breed (Nessah, 2020).

	Height (cm)	Weight (kg)
Male	66	60
Female	55	47

Horning: horns pointing backwards.

Coat: The coat is polychrome, with the dominant colours being beige, red, white, red piebald, black piebald and black. The coat can be long (3-9 cm in 46% of cases) or short (less than 3 cm in 54% of cases).

Ability: Its milk production is poor, but it is generally bred for meat production, which is of appreciable quality (Boubekeur, 2010; Laouadi, 2020, Nessah, 2020, Wikipedia, 2022).

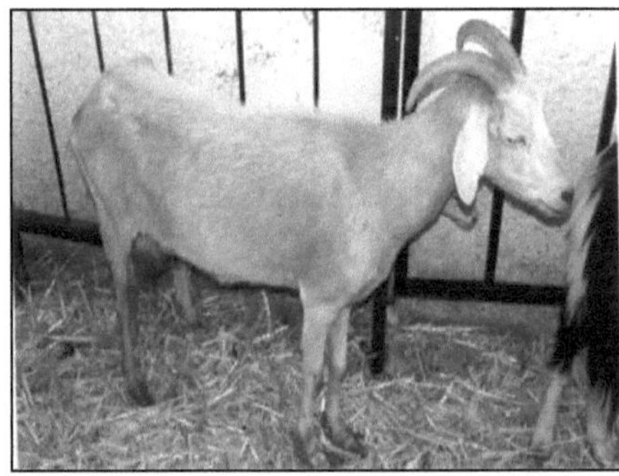

Figure 42: The Naine de Kabylie goat (Itelv, 2015).

VII.1.4. The M'ZABIT goat :

Also known as "the red goat of the oases". It originates from Metlili or Berriane.

Breed characteristics :

Morphology: reduced size (ellipometric). Body elongated, straight and rectilinear.

Head: convex profile. Thin head. The ears are long and hanging (15cm).

Table 39: Size and weight of the Mozabite breed (Itelv, 2015).

	Height (cm)	Weight (kg)
Male	68	50
Female	65	35

Horning: horns set back when present.

Coat: The coat is of three colours: buff, which dominates, brown and black. The hair is short (3-7cm) on most individuals.

Ability: this breed is very interesting in terms of milk production (2.56 kg/d) (Boubekeur, 2010; Laouadi, 2020, Nessah, 2020, Wikipedia, 2022).

Figure 43: The Mozabite goat (Itelv, 2015).

VII.2 The cross-over population :

Made up of subjects from controlled or uncontrolled crosses of local breeds with imported breeds (Maltese, Damasquine, Murciana, Toggenburg, Alpine and Saanen). The purpose of these crosses varies from region to region and from breeder to breeder (Boubekeur, 2010; Laouadi, 2020, Nessah, 2020, Wikipedia, 2022).

VII.3. Improved breeds :

These breeds have been introduced into Algeria since the colonial period, as part of a strategy of genetic improvement of the goat herd. They are Maltese, Murciana, Toggenburg and more recently Alpine and Saanen (Boubekeur, 2010; Laouadi, 2020, Nessah, 2020, Wikipedia, 2022).

- Maltese and Murcian goats were introduced to Oran and the coast during colonisation.

- The Maltese is found in the coastal areas of Annaba, Skikda, Algiers and the oases.

List of references

AMADOU TCHOUSSOU ISSAKA. Evaluation of serum biochemical parameters in local breed cows. Master's thesis, Université 8 Mai 1945 Guelma, 77 pages.

BABO DANIEL. Races Bovines Françaises - Editions France Agricole, 1st Edition 1998.

BABO DANIEL. Races Ovines et Caprines Françaises - Editions France Agricole, 2000.

BAKER FIONA. Running A Small Beef Herd - Landlinks Press, Third Edition, 162 pages, 2008.

BLANCHIN JEAN YVES. Le logement du mouton, élevages allaitants - Institut d'élevage, éditions France Agricole, 2005.

BOUBEKEUR ABDERRAHMANE 2010. Essai d'établissement de typologies d'exploitations d'élevages laitiers dans le contexte du Sud Algérien. 2010.

BOUSHABA NADJAT. Contribution à la caractérisation des races ovines algériennes, marocaines et françaises par l'utilisation des microsatellites OarFCB128 et l'étude des leurs relations phéylogénétiques. Magister thesis, University of Oran ES Senia, 2007, 119 pages.

DE WECKHERLIN D'AUGUSTE. Zootechnie Générale Des Animaux Domestiques - Libraire Victor Masson, 1857.

DERVILLE Marie, PATIN STEPHANE, AVON LAURENT - Races Bovines De France - Edition France Agricole, 269 pages, 2009.

DIRAND ANDRE. L'élevage du mouton - éditions Educagri, 2007.

FAUVE JEAN MICHEL, LE NEINDRE PIERRE. Ethologie appliquée - Editions Quae, pages 56-66, 2009.

FELIACHI. Rapport national sur les ressources génétiques animales: algérie commission nationale angr, **2003**.

FOURNIER ALAIN. L'élevage des chèvres - ARTEMIS éditions, 2006.

FOURNIER ALAIN. L'élevage des moutons - ARTEMIS éditions, 2006.

GILLESPIE De JAMES R., FLANDERS FRANK. Modern Livestock & Poultry Production - 8th edition, DELMAR CENGAGE Learning, Copyright 2009.

Institut Technique des Elevage (ITELV). Guide d'élevage des caprin. Baba Ali, 2015.

JUSSIAU R., PAPET A., RIGAL J., ZANCHI E. Amélioration Génétique Des Animaux D'élevage (Génome, Caractères, Sélection Et Croisements) - Educagri Editions, Dijon, 2013.

LAKHDARI FATTOUM. Guide de caractérisation phenotypique des races ovines de l'Algérie - Edition CRSTRA, 2015.

LAOUADI MOURAD. Caractérisation Et Gestion Des Principales Ressources Génétiques Caprines Locales En Algérie : Cas De La Région De Laghouat. PhD thesis, ENSV ? 250 pages. 2020

MERCIER EMMANUEL. Drawings of cattle breeds, Larousse Archives 2022 - www.larousse.fr .

NESSAH KAHINA. Characterisation zootechniques et génétique de la race caprine "naine kabyle" à Tizi Ouzou. PhD thesis, ENSV, Algiers. 2020.

PETTER F. Origine, modalités et conséquences de la domestication - Bull. Acad. Vet. De France, 423-427, 1987.

VAISSAIRE Jean-Pierre. Mémento De Zootechnie - Editions France Agricole, 252 Pages, 2014.

Yahimi ABDELKRIM. Some morphometric and reproductive characteristics of Atlas Brown bulls in Algeria. Revue d'élevage et de médecine vétérinaire des pays tropicaux, 74(2), p. 127-134, 2021.

WEBSITES :

Algerian sheep breeds: http://www.algerlablanche.com/index.php?post/Races-ovines-d-Alg%C3%A9rie

KEBBAB SALIM. Algerian sheep breeds: a genuine heritage: https://www.elwatan.com/edition/contributions/les-races-ovines-algeriennes-un-veritable-patrimoine-31-08-2018

List and classification of sheep breeds in France: https://www.doc-developpement-durable.org/file/Elevages/Moutons-Ovins/races/Liste%20et%20classification%20des%20races%20ovines%20de%20France_Wikipedia-Fr.pdf

MEYER ed. Dictionary of Animal Sciences. [On line]. Montpellier, France, CIRAD. [19/05/2022]. URL: http://dico-sciences-animales.cirad.fr/ ≥ 2022

Printed by Books on Demand GmbH, Norderstedt / Germany